The Facts on Matter

What are SOLUTIONS?

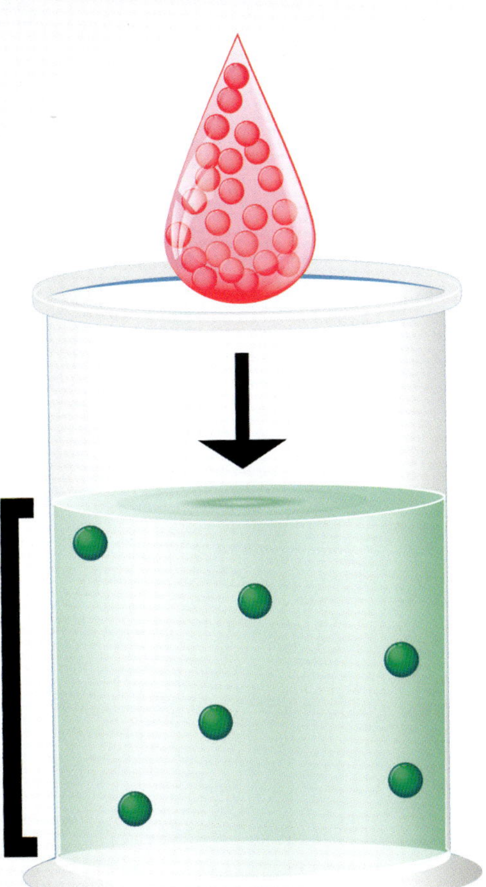

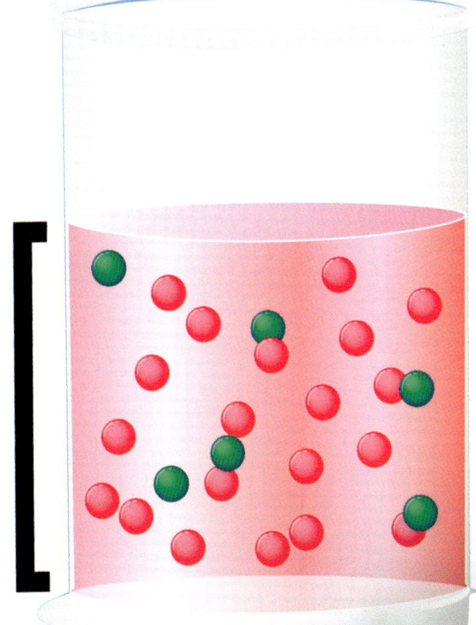

By Elise Tobler

Please visit our website, www.garethstevens.com. For a free color catalog of all our high-quality books, call toll free 1-800-542-2595 or fax 1-877-542-2596.

Library of Congress Cataloging-in-Publication Data

Names: Tobler, Elise, 1970- author.
Title: What are solutions? / Elise Tobler.
Description: New York : Gareth Stevens Publishing, [2022] | Series: The facts on matter | Includes index.
Identifiers: LCCN 2020032263 (print) | LCCN 2020032264 (ebook) | ISBN 9781538267134 (library binding) | ISBN 9781538267110 (paperback) | ISBN 9781538267127 (set) | ISBN 9781538267141 (ebook)
Subjects: LCSH: Solution (Chemistry)–Juvenile literature. | Mixtures–Juvenile literature.
Classification: LCC QD541 .T623 2022 (print) | LCC QD541 (ebook) | DDC 541/.34–dc23
LC record available at https://lccn.loc.gov/2020032263
LC ebook record available at https://lccn.loc.gov/2020032264

First Edition

Published in 2022 by
Gareth Stevens Publishing
29 E. 21st Street
New York, NY 10010

Designer: Katelyn E. Reynolds
Editor: Char Light

Photo credits: Cover, pp. 1 (solution), 6, 16 ttsz/ iStock / Getty Images Plus; cover, pp. 1–24 (background) sudanmas/E+/ Getty Images; cover, pp. 1–24 (chemistry doodles) backUp/Shutterstock.com; cover, pp. 1–24 (banner) Ozz Design/ Shutterstock.com; cover, pp. 1–24 (paper) Sergey Mironov /Shutterstock.com; p. 5 VectorMine/Shutterstock.com; pp. 7, 13 JGI/Jamie Grill/Getty Images; p. 9 Tom Le Goff/ DigitalVision/Getty Images; p. 11 krystiannawrocki/ E+/Getty Images; p. 15 (top) joshblake/ E+/Getty Images; p. 15 (bottom) Eternity in an Instant/ DigitalVision/Getty Images; p. 17 Tony Robins/ Photolibrary / Getty Images Plus; p. 19 Monica Fecke/ Moment Open/Getty Images; p. 21 Mel Yates/ The Image Bank / Getty Images Plus.

Printed in the United States of America

CPSIA compliance information: Batch #CWGS22: For further information contact Gareth Stevens, New York, New York at 1-800-542-2595.

CONTENTS

Words in the glossary appear in **bold** type the first time they are used in the text.

Properties of SOLUTIONS

A solution is a mixture of two or more **substances**. One of the substances **dissolves** into the other. Once dissolved, you cannot see the individual components, or parts, on their own. Solutions are so **dense** that they will not **scatter** a light beam if you try to shine one through!

In a solution, the individual parts of a solution do not separate or **settle**. They stay **blended**. However, you can still separate solutions into their components.

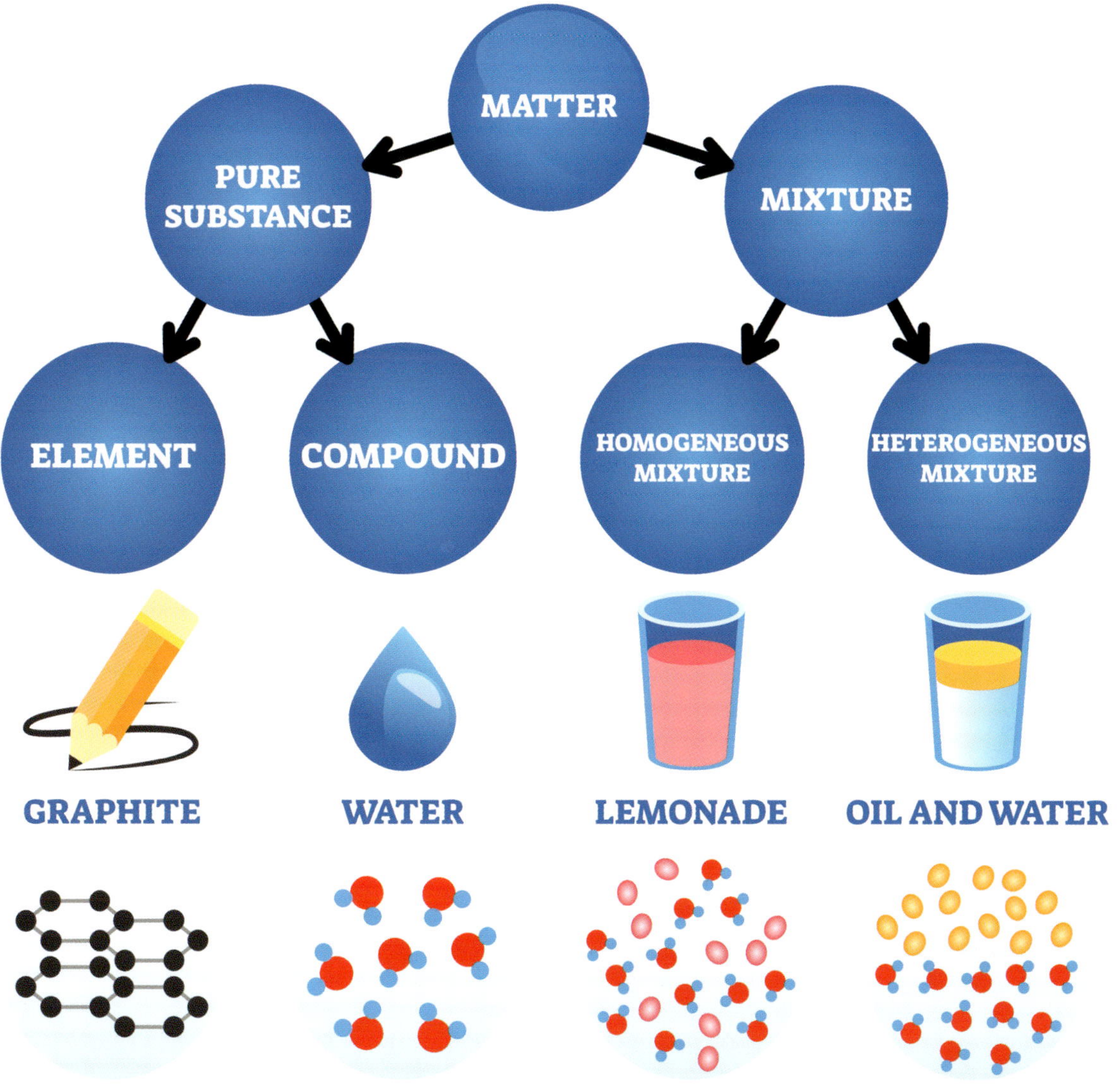

Solutions are a kind of **homogeneous** mixture, like lemonade!

Solutes and SOLVENTS

Every solution has two parts—the solute and the solvent. The solute is the substance being dissolved. The solvent is the liquid the solute is dissolving into. When cream is stirred into hot coffee, the cream is the solute and the coffee is the solvent.

Water is considered the "**universal** solvent," because it dissolves most **chemicals**. When a solution cannot hold anymore solute, it's said to be saturated. When a solution contains too much solute, it's supersaturated.

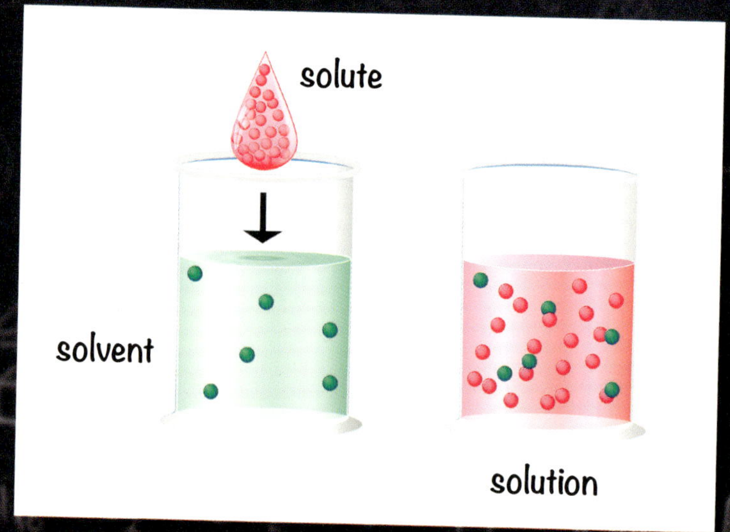

Sometimes, a solute needs to be mixed with the solvent in order to evenly spread. Sugar needs to be stirred to mix evenly with coffee. Once it's evenly spread, or homogeneous, it's a solution.

SOLUBILITY

Solubility means the ability of a solute to dissolve into a solvent. We can measure solubility as the highest amount of solute that is able to dissolve in a given amount of a solvent at a certain temperature.

If you have one cup of salt, and you add one spoonful of water, you haven't made a solution. There isn't enough water (solvent) for all of the salt (solute) to dissolve! But if you add the salt to a bucket of water, the salt will dissolve. You made a solution!

Know the Basics!

Solutions boil at higher temperatures and freeze at lower temperatures than the components would if they were separate.

When something is water-soluble, that means it dissolves in water!

Solid SOLUTIONS

Solutions can contain two different states of matter, but one always dissolves into the other. Sterling silver is a solution. It is mostly silver, but copper is mixed in to make it stronger. These combinations of metals are called alloys.

Alloys have different properties than their individual metals. Alloys do not have a single **melting point**, but a range of them. Coins are also considered solutions. The pennies in your pocket are an alloy of copper and zinc.

Solder, which is used in computers, is an alloy and also a solution.

Liquid SOLUTIONS

Liquid solutions can be made up of two liquids or a liquid and solid. An example of a liquid-liquid solution is gasoline. An example of a liquid-solid solution would be the substance dentists use to fill cavities in teeth.

Pickles are made with a liquid solution. Salt and sugar are dissolved in vinegar. Cucumbers and dill are added. The liquid solution preserves, or pickles, the cucumbers so they will keep longer than cucumbers do on their own.

Know the Basics!

Milk is a special solution, called a colloid. Colloids are solutions with larger particles of solute, and they are usually cloudy or milky.

If you have ever stirred chocolate syrup into milk, you have made a solution! Chocolate syrup is also a solution on its own, before it becomes chocolate milk.

Gas SOLUTIONS

Gas solutions often include two states of matter, but gas-gas solutions are also possible. Air is a gas-gas solution, including several kinds of gases. You can also find liquid-gas solutions. For example, drops of water (liquid) in the air (gas) make fog.

Soda is also a liquid-gas solution. When soda is sealed in a can, you can't tell the gas bubbles apart from the liquid soda. They are both smoothly blended. Only when the can is opened does the gas fizz, or hiss and bubble.

Know the Basics!

Body fluids like blood are considered liquid solutions. When they change, it can be a sign to doctors that someone is ill or hurt.

Cream combines with the gas nitrous oxide to instantly become whipped cream, a liquid-gas solution.

MISCIBILITY

Some liquids can form solutions and others cannot. Vinegar and water blend smoothly. When this happens, we say that the liquids are miscible. When liquids don't easily blend, like oil and water, they are called immiscible.

When oil is mistakenly spilled in the ocean, the oil floats on the water and does not blend in. Some oil and water combinations can be forced to blend together, like in salad dressings, but only with a lot of shaking or stirring.

Depending on their properties, some liquids won't blend into each other and pile on top of one another instead.

Separating SOLUTIONS

Unlike mixtures, solutions cannot usually be separated by simple **filters**, unless they contain a solid that is insoluble, or unable to dissolve. To separate a solution, one can try evaporation, distillation, or chromatography.

Evaporation, where a liquid turns to a vapor (gas), doesn't usually require heat. Distillation does need heat. It uses the different **boiling points** of the components. When one boils, the other is left behind. Chromatography passes evaporating vapor through paper to see its components.

Know the Basics!

Hot water has more energy, or power, than cold water. Hot liquids can hold more solute than cold because the solute can break apart more easily under heat.

Making rock candy is one way to separate sugar from water!

Concentrates and DILUTES

Concentrated solutions have more solute than solvent. Liquid hand soap and cough syrup are concentrated solutions. Cans of soup are also sometimes concentrated. When you have to add a can of water to the soup, you are diluting its concentrated form, or thinning it out.

Dilute solutions contain less solute. Tap water is an example of a dilute solution, because the water makes up most of the mixture, with fewer dissolved solids.

Know the Basics!

The saturation point is the point of the highest concentration of a solute in a solvent. This point depends on temperature and the chemical nature of the solution.

The darker a solution is, the more solute it contains.

GLOSSARY

blend: to thoroughly combine

boiling point: the temperature at which matter boils

chemical: a substance that is a product of a chemical process

dense: packed very closely together

dissolve: to be mixed into a liquid so that it forms a solution

filter: a tool that is used to remove something unwanted from a liquid or gas that passes through it

homogeneous: having to do with a mixture that has all parts evenly spread throughout it

melting point: the temperature at which a solid will melt

scatter: to break apart

settle: to fall to the bottom level of a container

substance: a certain kind of matter

universal: applying to all cases

For More INFORMATION

Books

Barfield, Mike. *The Element in the Room: Investigating the Atomic Ingredients That Make Up Your Home*. London, England: Laurence King Publishing, 2017.

Chatterton, Crystal. *Awesome Science Experiments For Kids*. Berkeley, CA: Rockridge Press, 2018.

Silver, Donald; Wynne, Patricia T. *My First Book About Chemistry*. Mineola, NY: Dover Publications, 2020.

Websites

Chem4Kids
www.chem4kids.com/
This site contains everything you need to know about chemistry!

Ducksters
www.ducksters.com/science/chemistry/solutions_and_dissolving.php
Learn more about solutions of all kinds.

ScienceFun
www.sciencefun.org/kidszone/experiments/
The Kids Zone contains easy-to-do science experiments!

INDEX